My First Book Of Patterns
Numbers
1 TO 20

Wonder House

1 one

Count and color the only sun in the sky.

Trace the number

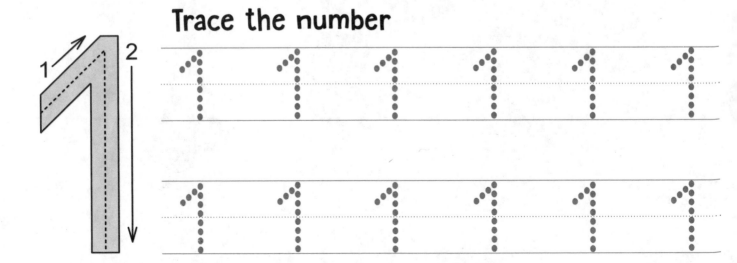

2 two

Count and color the tyres of the bicycle.

Trace the number

3 three

Count and color the lotus flowers in the pond.

Trace the number

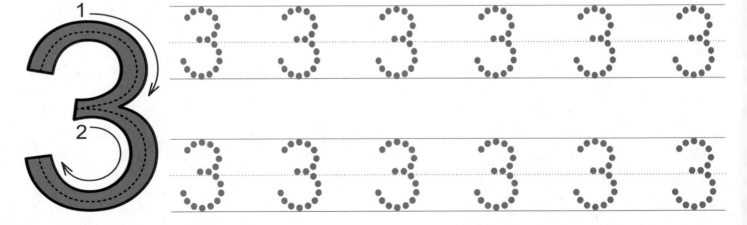

4 four

Count and color the bird houses.

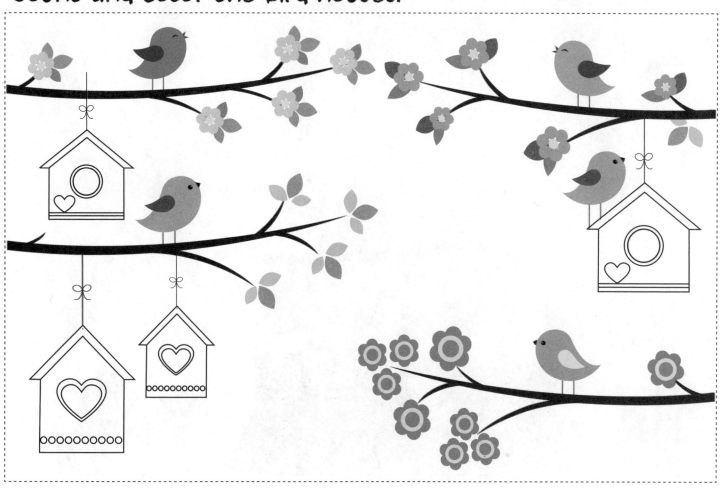

Trace the number

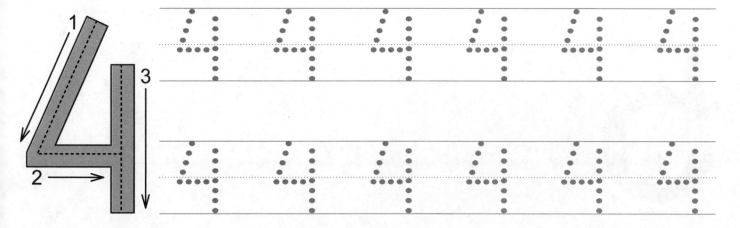

5 five

Count and color the candles on the cake.

Trace the number

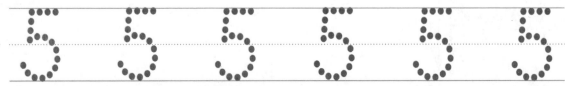

Match Them!

Trace the numbers and match with the correct group of items.

In the Farm

Count the things you see in the farm. Write their correct numbers in the boxes given below with their pictures.

A

B

C

D

E

F

G

H

I

J

Farm Count

Count and write the number of items in each box.

A.

B.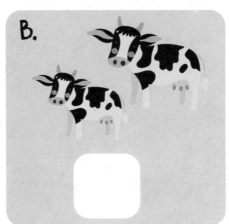

C.

D.

E.

F.

G.

H.

I.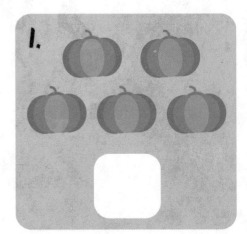

6 six

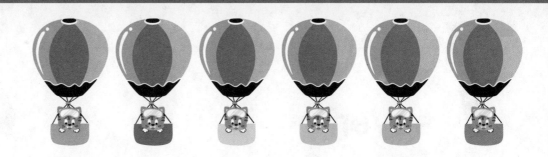

Count and color the air balloons in the sky.

Trace the number

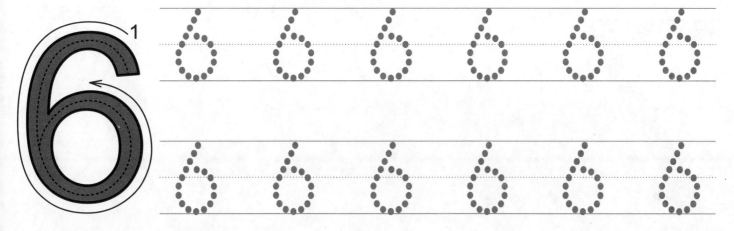

7 seven

Count and fill the colors in the rainbow.

Trace the number

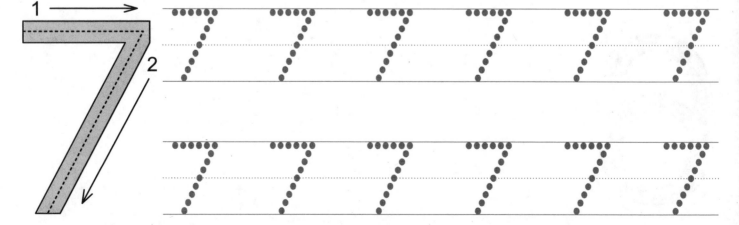

8 eight

Trace the numbers on the carts of the giant wheel.

Trace the number

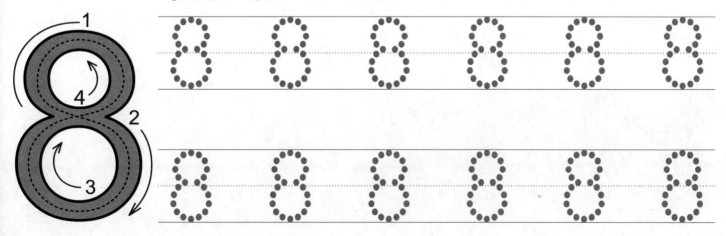

9 nine

Count and color the ducks in the pond.

Trace the number

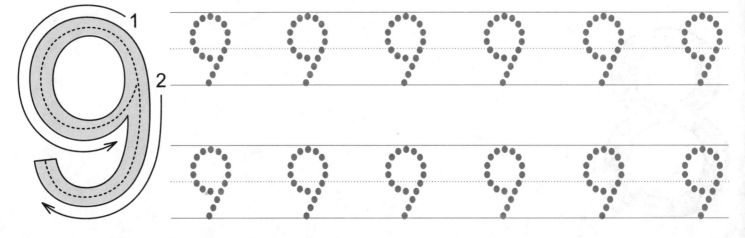

10 ten

Count and color the lollipops.

Trace the number

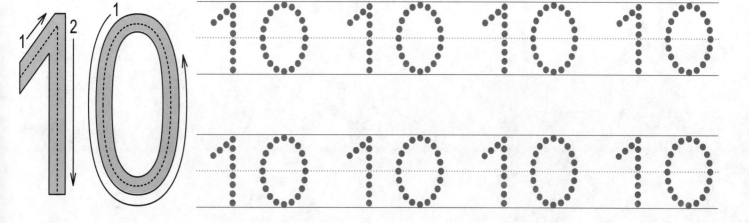

At the Zoo

Count the animals you see at the Zoo. Write their numbers in the boxes given below with their picture.

A

B

C

D

E

F G H I J

Zoo Count

Count and write the number of animals in each box.

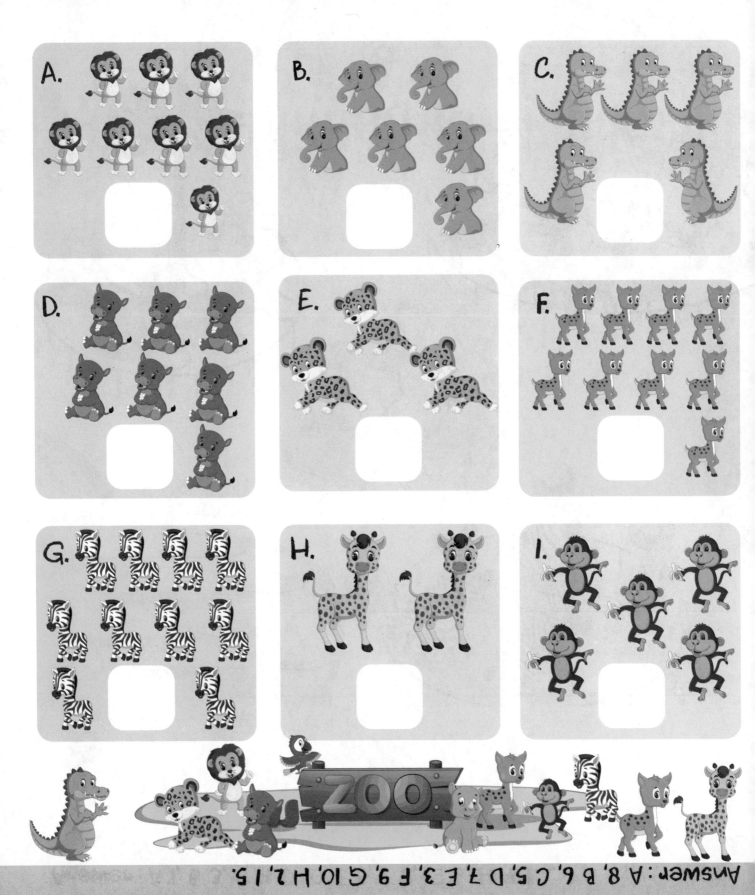

Match Them!

Trace the numbers and match with the correct group of items.

Say Aloud!

Trace the number on each tractor and say their names aloud.

Missing Numbers!

Fill in the blanks to complete the number sequence.

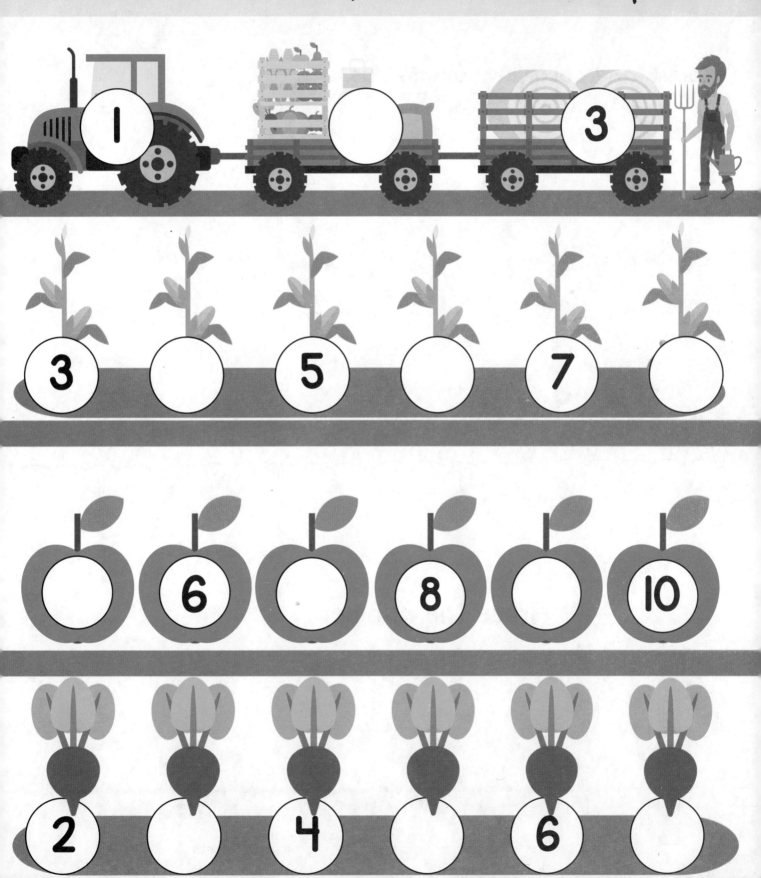

11 eleven

Count and color the umbrellas.

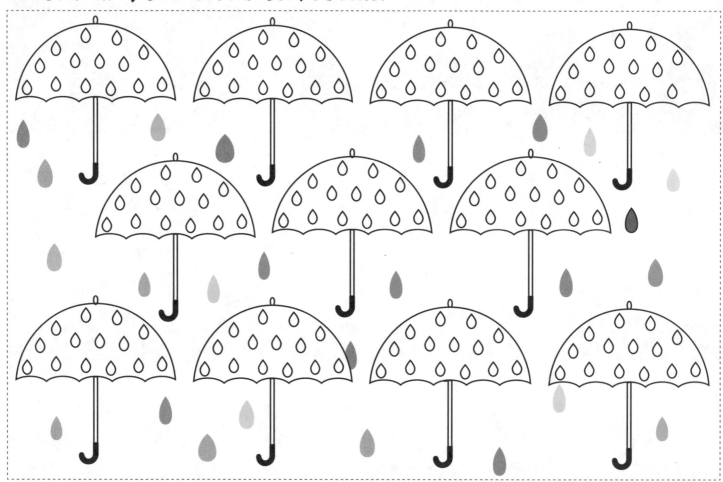

Trace the number

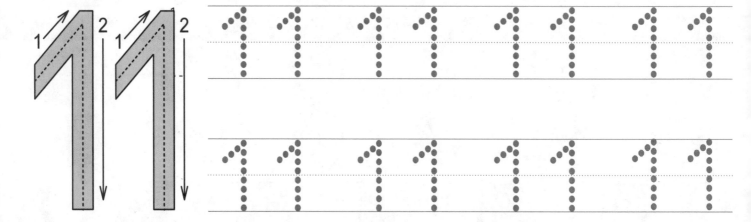

12 twelve

Count and color the ants walking towards the ant hill.

Trace the number

13 thirteen

Count and color the balloons.

Trace the number

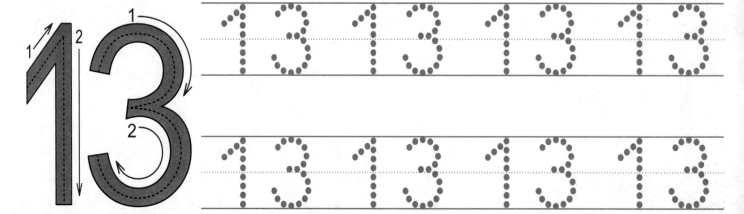

14
fourteen

Count and color the butterflies in the garden.

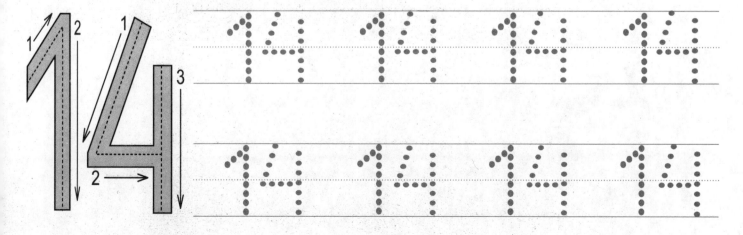

Trace the number

14

15
fifteen

Count and color the ice creams.

Trace the number

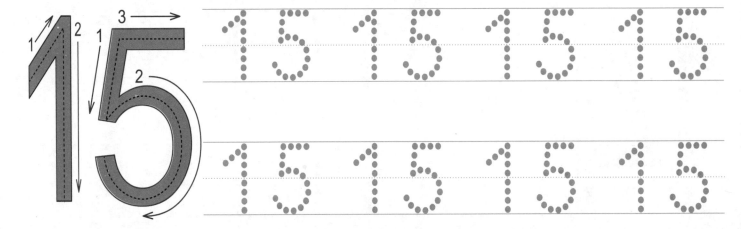

Match Them!

Trace the numbers and match with the correct group of items.

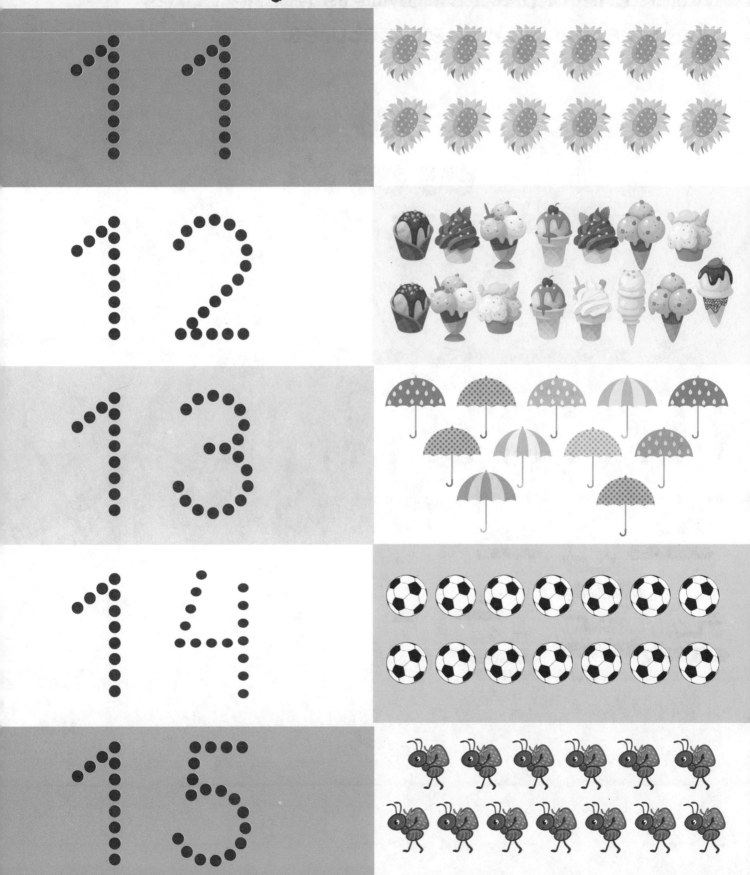

Amusement Park

Count the things you see in the amusement park.
Write their correct numbers in the boxes
given below with their pictures.

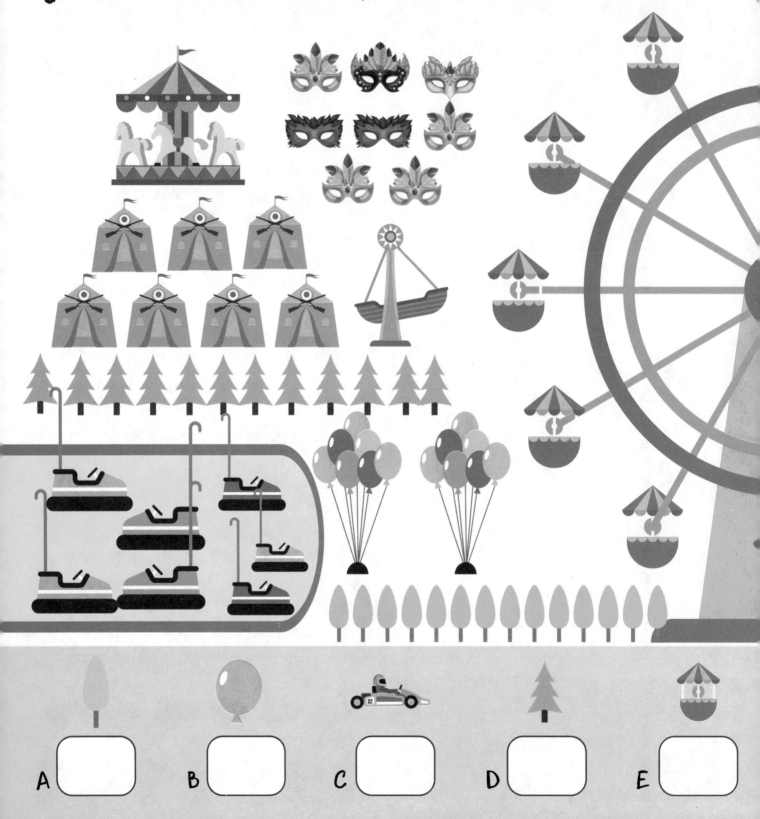

A

B

C

D

E

F ⬜ G ⬜ H ⬜ I ⬜ J ⬜

Go-Karting Track

Rank the winners of the go-karting competition.

First

Second

According to their position in the competition
Write the car number in the boxes given below.

Third

Fourth

Fifth

16 sixteen

Count and color the ladybirds.

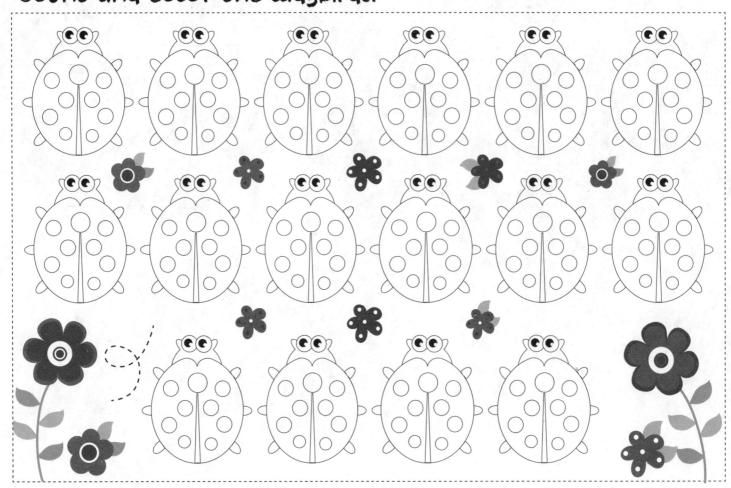

Trace the number

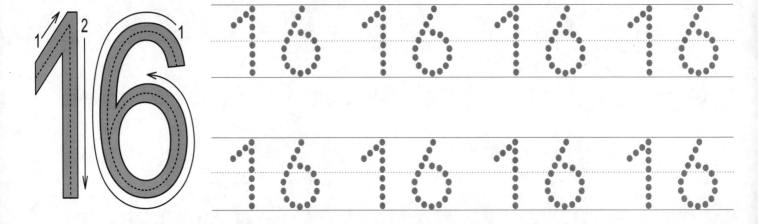

17 seventeen

Count and color the flowers.

Trace the number

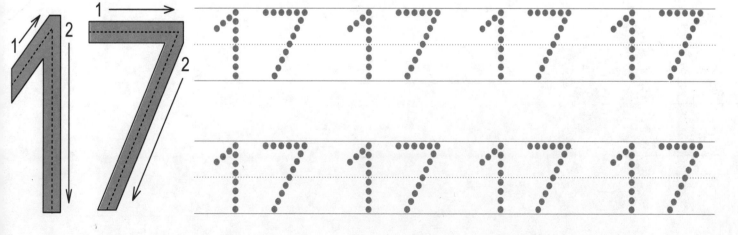

18

eighteen

Count and color the vehicles on the street.

Trace the number

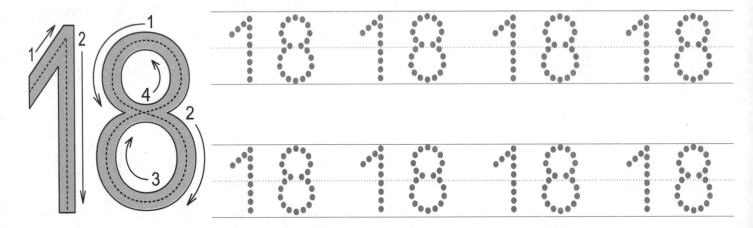

19
nineteen

Count and color the bees.

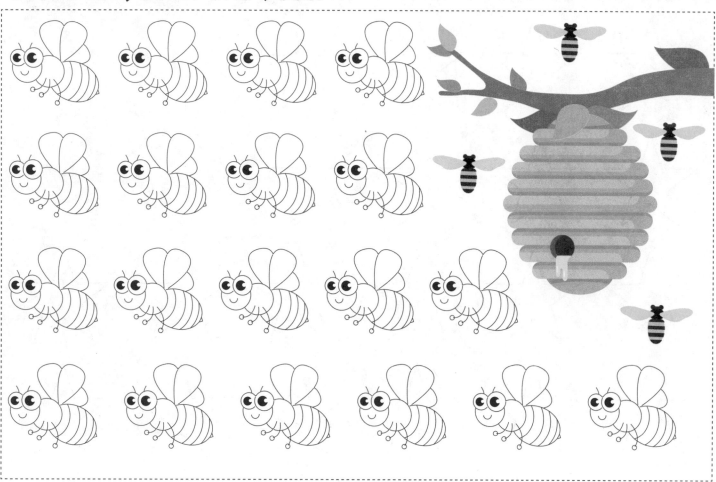

Trace the number

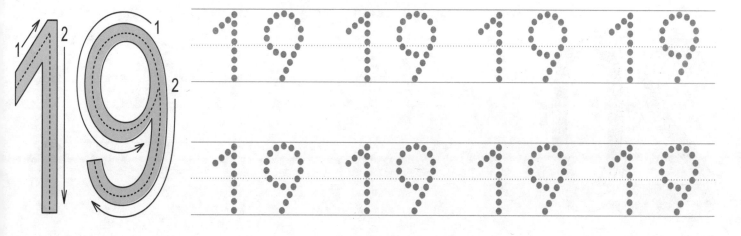

20 twenty

Count and color the stars in the sky.

Trace the number

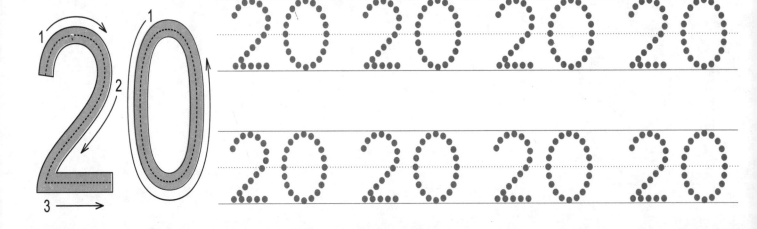

Match Them!

Trace the numbers and match with the correct group of items.

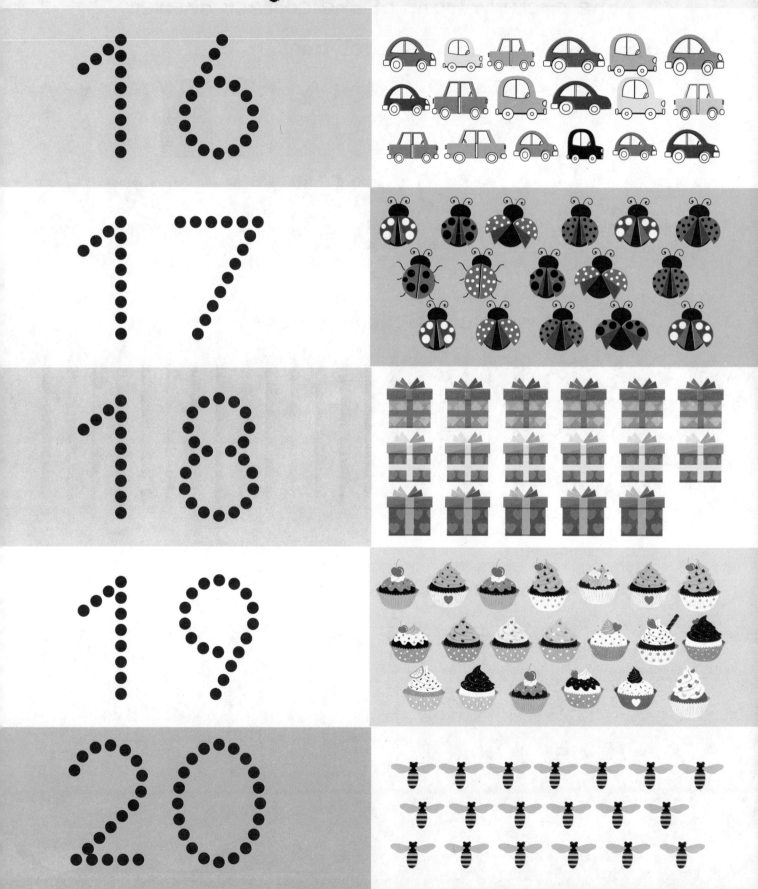

Picture Counting

Count the number of items in each box and color
the circle in front of the correct answer.

1. Count the number of school bags.

- ◯ A 16
- ◯ B 17
- ◯ C 19

2. Count the number of color pencils.

- ◯ A 16
- ◯ B 20
- ◯ C 19

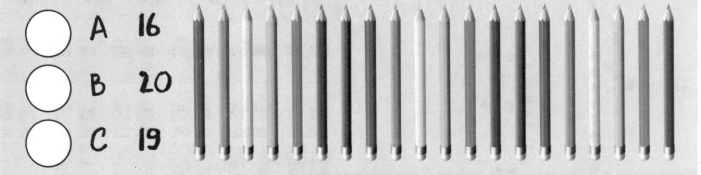

3. Count the number of paint brushes.

- ◯ A 16
- ◯ B 17
- ◯ C 19

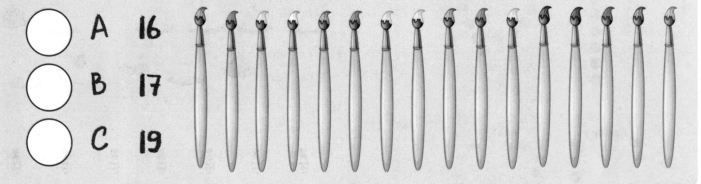

4. Count the number of color palettes.

○ A 15

○ B 17

○ C 19

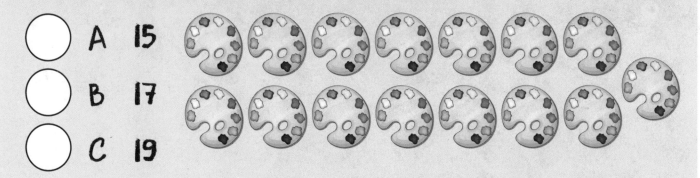

5. Count the number of Water bottles.

○ A 16

○ B 17

○ C 18

6. Count the number of Lunch boxes.

○ A 16

○ B 17

○ C 19

Answer : 1A, 2B, 3A, 4A, 5C, 6B.

In the Ocean

Count the things you see under the ocean. Write their correct numbers in the boxes given below with their picture.

A ☐   B ☐   C ☐   D ☐   E ☐

F

G

H

I

J

Treasure Hunt

Write the missing numbers from 1 to 20 and help the pirate find treasure by following the route.

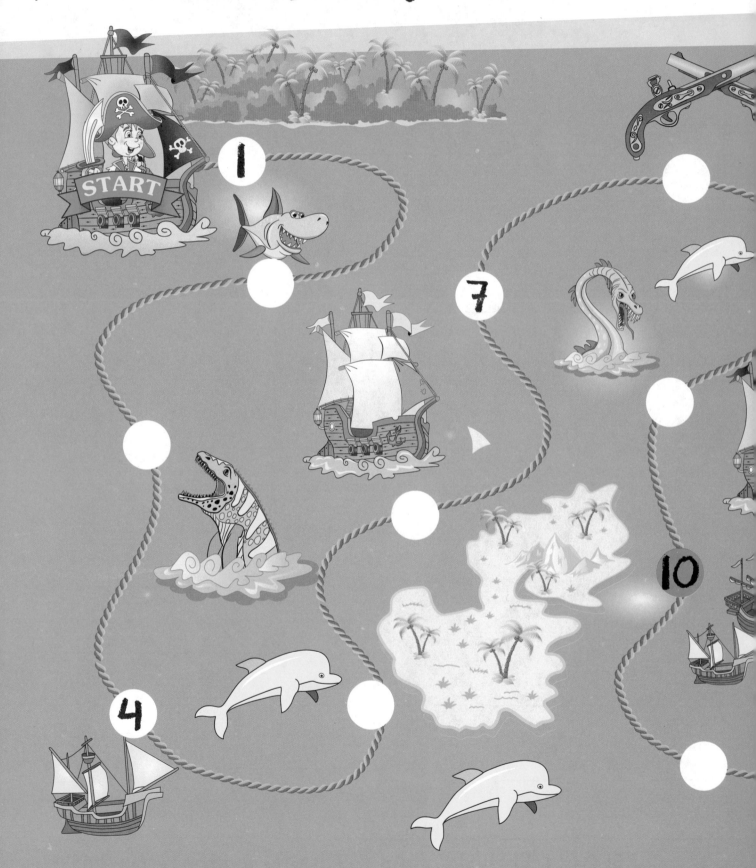

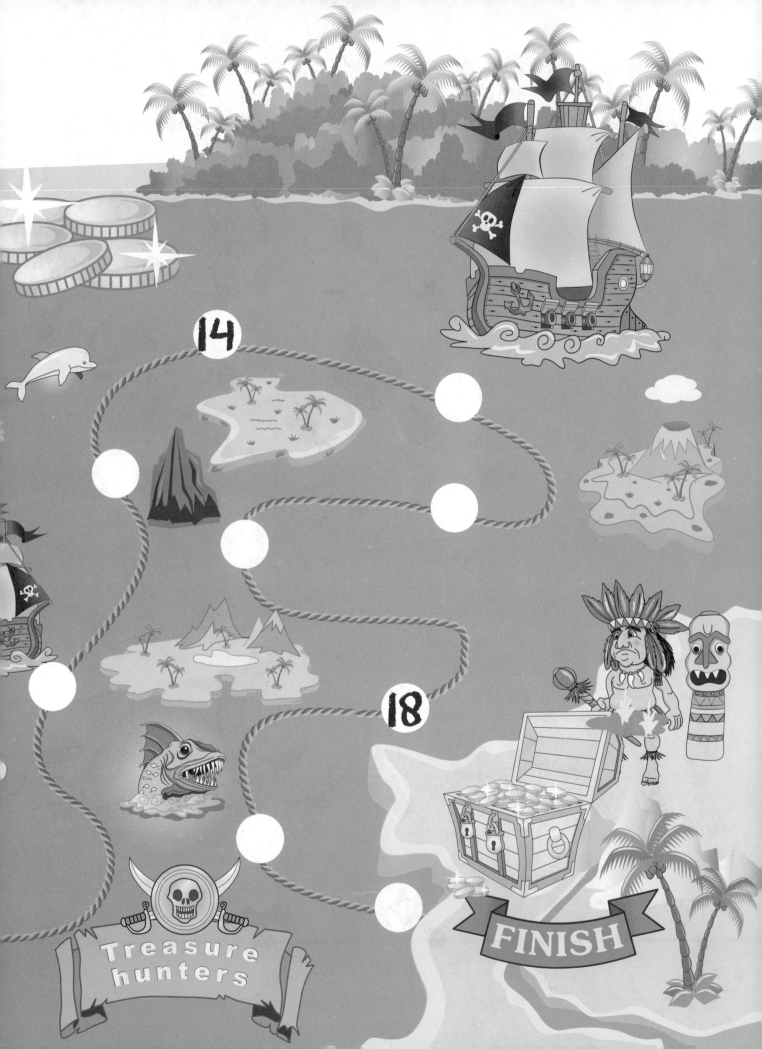

14

18

Treasure
hunters

FINISH

On the Tree

Count the birds you see on the tree. Write the correct number of birds in the boxes given below their pictures.

A [ ]

B [ ]

C [ ]

D [ ]

E [ ]

Answer : A 10, B 16, C 6, D 14, E 9, F 8, G 12, H 13, I 11, J 15.

In the Kitchen Garden

Count the things you see in the kitchen garden.
Write their correct numbers in the boxes given below
their pictures.

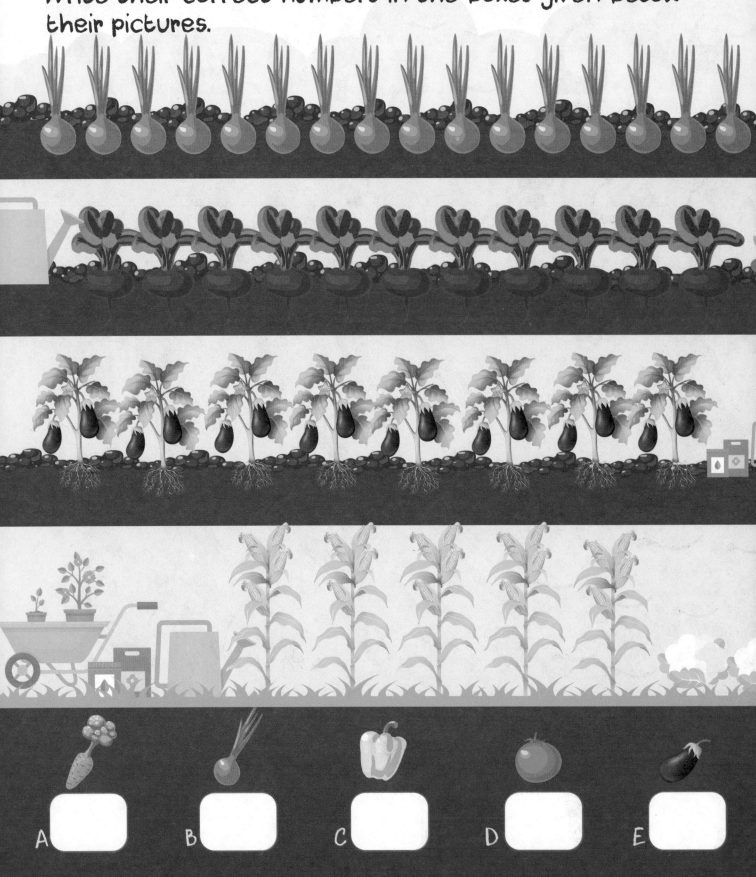

A

B

C

D

E

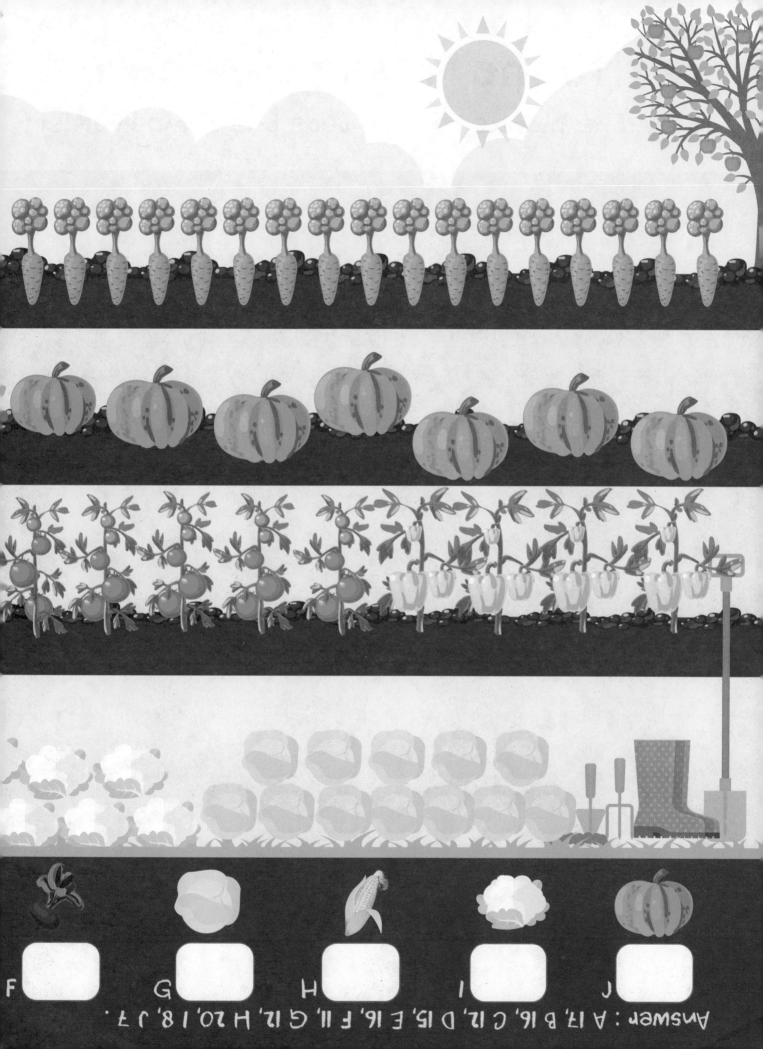

F ▢   G ▢   H ▢   I ▢   J ▢

Answer : A.17, B.16, C.12, D.15, E.16, F.11, G.12, H.20, I.8, J.7.

Write the Missing Numbers.

Fill in the blanks to complete the number order.

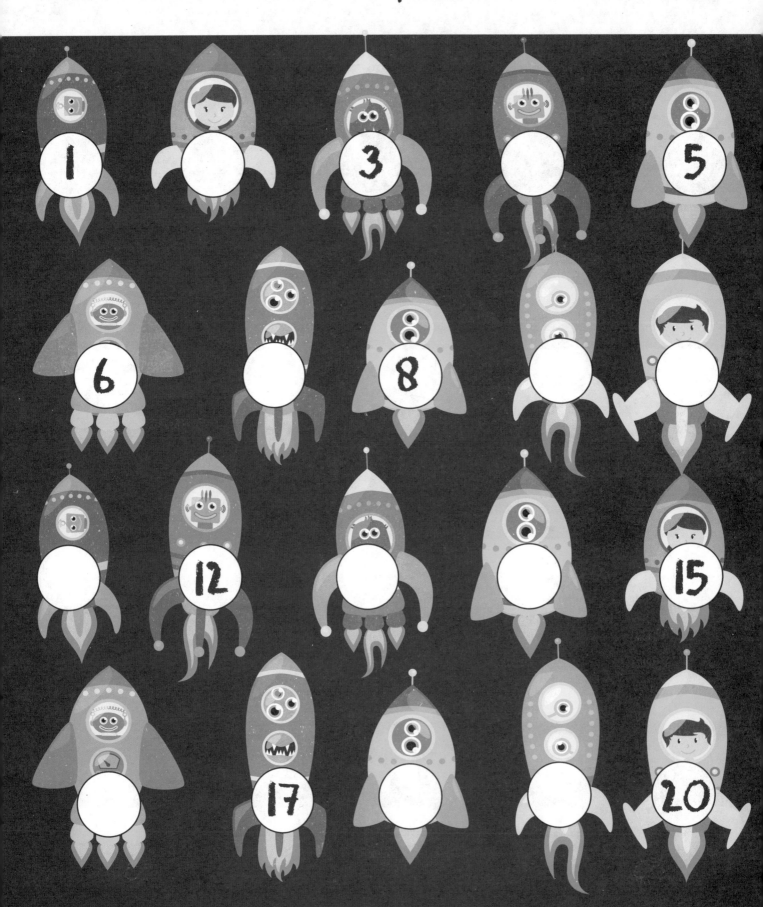